STUDYING SECRETIVE ANIMALS IN THE WILD

GIANT ARMADILLOS

by Joyce Markovics

Cherry Lake Press

Published in the United States of America by Cherry Lake Publishing Group
Ann Arbor, Michigan
www.cherrylakepublishing.com

Reading Adviser: Marla Conn, MS Ed., Literacy specialist, Read-Ability, Inc.
Content Adviser: Arnaud Desbiez
Book Designer: Ed Morgan

Photo Credits: © Kevin Schafer, cover and title page; © freepik.com, TOC; © mariamalaya/Shutterstock, 4-5; © Arnaud Desbiez, 5 left and right; © Tomaz Nascimento de Melo, 6; © Arnaud Desbiez, 7; Wikimedia Commons, 8; Wikimedia Commons, 9; © belizar/Shutterstock, 10; © Arnaud Desbiez, 11; © Arnaud Desbiez, 12; © Arnaud Desbiez, 13; freepik.com, 14-15; © Arnaud Desbiez, 14 bottom; © darock/Shutterstock, 15 top; © Uwe Bergwitz/Shutterstock, 15 bottom; © Kevin Schafer, 16; © Uwe Bergwitz/Shutterstock, 17 top; © begita/Shutterstock, 17 bottom; © Arnaud Desbiez, 18; © Arnaud Desbiez, 19; © Arnaud Desbiez, 20; © Arnaud Desbiez, 21; © Philip Lange/Shutterstock, 22; © Arnaud Desbiez, 23; © Arnaud Desbiez, 24; © Arnaud Desbiez, 25; © Arnaud Desbiez, 26-27; © Tomaz Nascimento de Melo, 28; © spatuletail/Shutterstock, 29; Prachaya Roekdeethaweesab/Shutterstock, 31.

Cherry Lake Press is an imprint of Cherry Lake Publishing Group.

Library of Congress Cataloging-in-Publication Data

Names: Markovics, Joyce L., author.
Title: Giant armadillos / by Joyce L. Markovics.
Description: First edition. | Ann Arbor, Michigan : Cherry Lake Publishing, [2021] | Series: On the trail: studying secretive animals in the wild | Includes bibliographical references and index. | Audience: Ages 10 | Audience: Grades 4-6
Identifiers: LCCN 2020030355 (print) | LCCN 2020030356 (ebook) | ISBN 9781534180468 (hardcover) | ISBN 9781534182172 (paperback) | ISBN 9781534183186 (ebook) | ISBN 9781534181472 (pdf)
Subjects: LCSH: Armadillos–Juvenile literature.
Classification: LCC QL737.E23 M323 2021 (print) | LCC QL737.E23 (ebook) | DDC 599.3/12–dc23
LC record available at https://lccn.loc.gov/2020030355
LC ebook record available at https://lccn.loc.gov/2020030356

Printed in the United States of America
Corporate Graphics

CONTENTS

A BIG FIND

It was nighttime in the Pantanal (pan-TEH-nal), a huge wetland in South America. Bugs chirped loudly under the glowing stars. **Field biologist** Arnaud Desbiez crouched next to a video screen watching and waiting. He was hoping to catch a **glimpse** of one of the rarest animals in the world—the giant armadillo.

LOOK CLOSER

The Pantanal is the biggest freshwater wetland in the world! It covers 80,000 square miles (207,199 square kilometers), including parts of Brazil, Paraguay, and Bolivia. It's about the same size as the state of Kansas.

After 10 years of living and working in the Pantanal, Arnaud had yet to spot one. "It was a **species** I really wanted to see," he says. Suddenly, Arnaud saw movement on the screen, which was connected to a special camera set up near a giant armadillo's **burrow**. Then, as if out of a dream, Arnaud saw the huge **armored** animal. It had claws the size of kitchen knives! Seeing the creature, Arnaud says, was "amazing."

This picture of a giant armadillo was taken at night with Arnaud's special camera.

Field biologist Arnaud Desbiez tracking giant armadillos

RARELY SEEN

Arnaud was astonished by the animal's size. Giant armadillos can grow up to 5 feet (1.5 meters) long and weigh over 100 pounds (45 kilograms)! But their large size has not made them easy to study. "The giant armadillo is rarely seen and poorly understood," says Arnaud. Why? Giant armadillos mostly live alone and are **nocturnal**. And they spend three-quarters of their lives underground in burrows.

In the Pantanal, local people once believed the giant armadillo was an underground monster.

Arnaud measures a giant armadillo's burrow. Often, the big burrow is the only sign of the animal.

In addition, giant armadillos are especially fearful of people. A strange scent or noise can send them scurrying into their burrows. Even locals who live close to the massive animals seldom see them. "It's right there. But you don't see it," says Arnaud.

LOOK CLOSER

Giant armadillos make their homes in large, deep burrows that can stretch 20 feet (6 m) into the earth!

ANCIENT ANIMALS

Armadillos belong to an ancient group of **mammals** called Xenarthrans (zuh-NAR-threnz). They first appeared on Earth about 80 million years ago. Around 2 million years ago, a relative of the giant armadillo roamed North and South America. It looked like a massive turtle with a long tail. This animal, called a glyptodont (GLIP-tuh-dont), grew up to 10 feet (3 m) long! That's about the size of a small car.

Glyptodonts fed mostly on plants.

LOOK CLOSER

Another ancient armadillo relative had a spiky tail, which it swung at enemies.

A shell-like **carapace** covered the glyptodont's body. Each carapace was made up of more than 1,000 bony plates. So what happened to these ancient animals? Over time, glyptodonts died out. Scientists aren't sure why. Luckily, says Arnaud, their giant armadillo relatives still live on.

A close-up view of a glyptodont's bony carapace

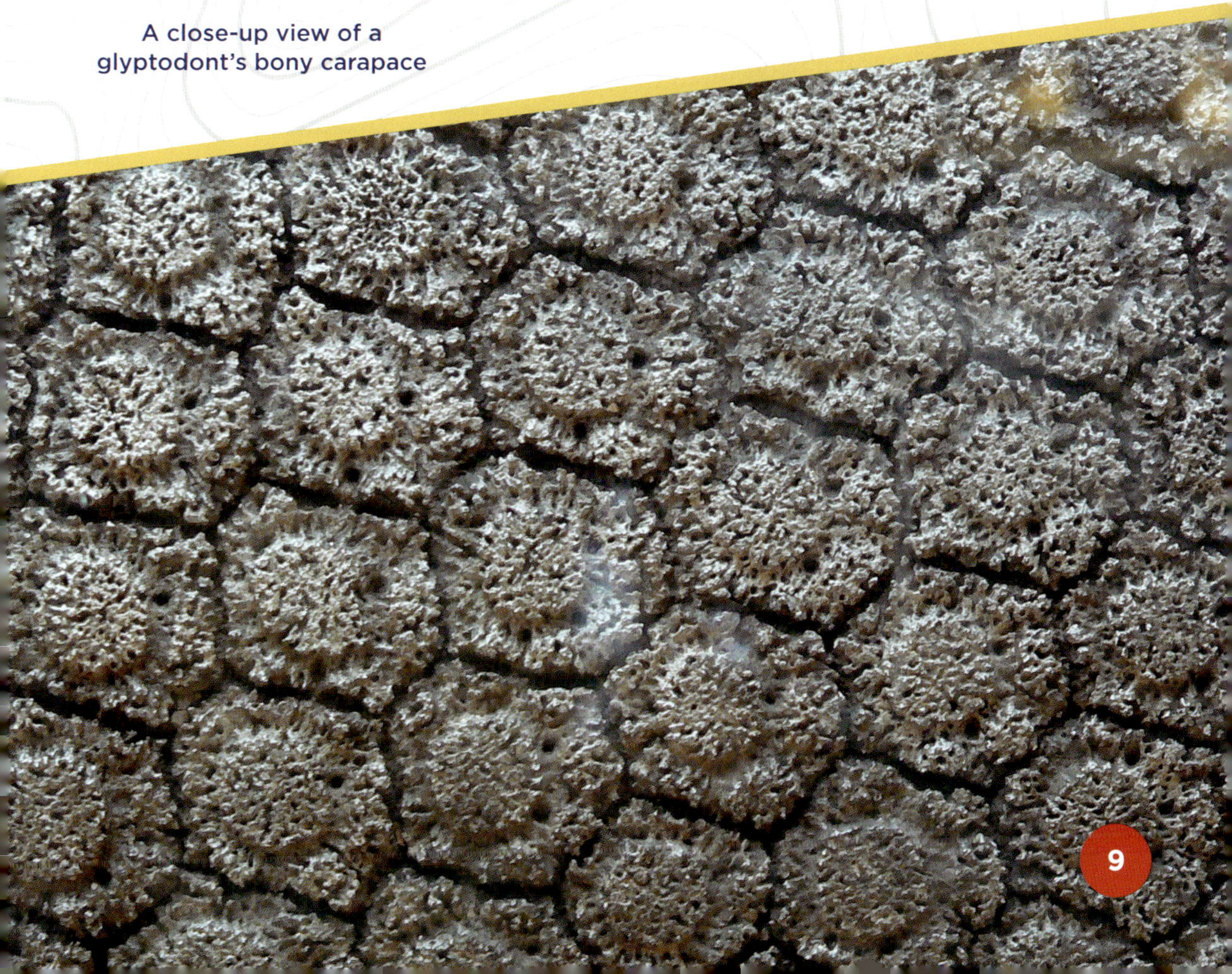

ARMORED TANKS

Today, there are 20 different kinds of armadillos. They range in size from fairy armadillos, which can fit in a child's hand, to giant armadillos. Similar to their ancient relatives, today's armadillos have a flexible carapace that's made of skin-covered bony plates called **scutes**. In fact, the word *armadillo* means "little armored one" in Spanish. Some species can roll up in a ball to stay safe from **predators**.

A southern three-banded armadillo rolled up in a ball

Because of their size, giant armadillos are unable to ball up, says Arnaud. However, their scutes cover their entire upper body and head. Thanks to their body armor, giant armadillos are well protected against enemies, such as jaguars and pumas, as well as prickly plants in the Pantanal.

The scutes of a giant armadillo

LOOK CLOSER

Armadillos are related to sloths and anteaters.

POWERFUL DIGGERS

After its body armor, "the second thing you notice when face-to-face with an armadillo is its claws," says Arnaud. Giant armadillos have huge front claws. They can grow up to 5.7 inches (14.5 centimeters) long—the largest of any mammal in the world! Even though they look like weapons, armadillos mainly use their claws for digging and finding food.

The claws of a giant armadillo

Their claws make it hard for giant armadillos to run away from threats. It's like "trying to **sprint** in high heels," says Arnaud. But giant armadillos are powerful diggers. When they put their claws to work, giant armadillos can bury themselves in just a few minutes. They can dig a burrow that's 1 foot (0.3 m) wide in about 20 minutes.

An armadillo uses its back legs to kick dirt out of the way when digging.

LOOK CLOSER

A giant armadillo's claws are made from the same material as human fingernails.

IMPORTANT BURROWS

Giant armadillos dig a new burrow about every 2 days—that's about 15 burrows a month. Using special cameras, Arnaud discovered that when the armadillos leave their burrows, dozens of other animals move in. These animals include tamanduas (tuh-MAN-doo-uhs), foxes, pigs, tegus (teh-GOOZ), agoutis (uh-GOO-tees), and jays—at least 70 different species altogether!

An agouti, a type of rodent, grabs a bite to eat in a giant armadillo's burrow.

Some of the animals use the armadillo burrows to escape the heat or cold, find safety, or raise their young. Others find food, like bugs or seeds, in the underground hideouts. Through digging, the armadillos shape and re-shape the land. That's why Arnaud calls giant armadillos **ecosystem engineers**.

An anteater called a tamandua usually lives in the trees. But it's also a frequent visitor to giant armadillo burrows.

A tegu, another burrow guest, is a kind of reptile.

LOOK CLOSER

When they dig, giant armadillos also help different plants grow by loosening the soil.

BUGS FOR DINNER

Giant armadillos leave their burrows at night to find a meal. They start sniffing the air in hopes of smelling ants and termites, their favorite foods. Even though they mainly eat insects, giant armadillos have more teeth than any other mammal—up to 100!

A giant armadillo sniffs the air.

Arnaud set up cameras on termite mounds, where the armadillos love to feed. A termite mound is made of dried dirt and termite spit and can be as hard as cement. Using their claws, giant armadillos can rip one apart in minutes! They have incredible strength, says Arnaud. In one feeding session, an armadillo can **devour** thousands of termites. After they're done eating, the giants leave behind holes in the termite mounds. This provides a ready-to-eat meal for other **insectivores** living in the Pantanal.

Termite mounds dot the Pantanal like small buildings. They can be up to 6.5 feet (2 m) tall!

A termite

LOOK **CLOSER**

An adult armadillo's teeth are small and peg-like. They lack enamel, which is the hard outer covering on most mammals' teeth.

CAPTURE AND RELEASE

In order to track and study giant armadillos, Arnaud first has to safely capture them. He does this by setting a trap at the opening of their burrows at night. Then he waits. After an animal is in the trap, Arnaud and his team of helpers rush to the burrow.

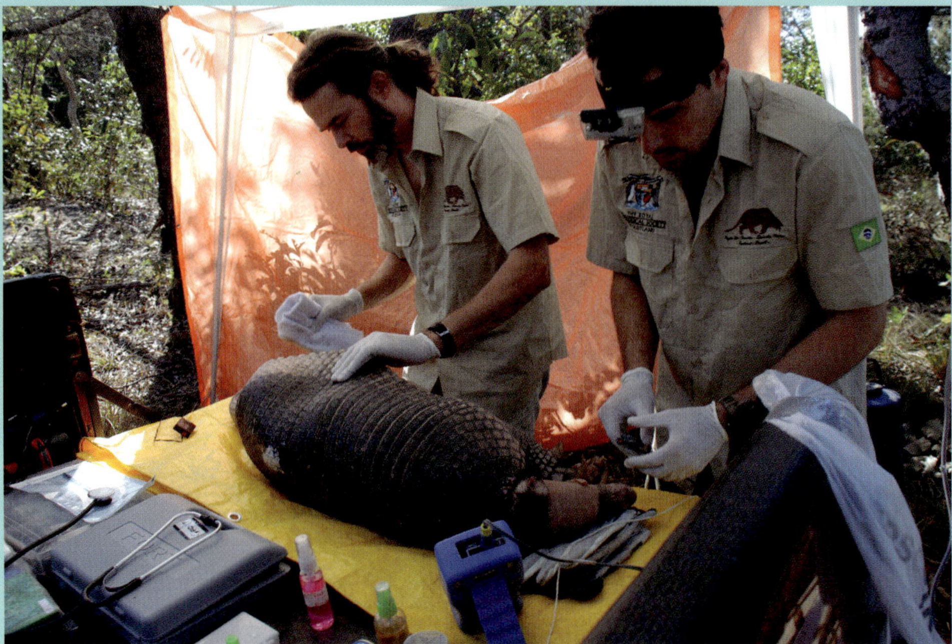

Arnaud and a team member examining an armadillo

Once captured, the giant armadillo is given a drug to make it sleepy. Then it's fitted with a tiny **transmitter**. This device allows Arnaud to find and track the creature at all times. After he tags the armadillo and safely wakes it back up, it's released back into its burrow.

A tagged giant armadillo runs back to its burrow.

LOOK CLOSER

Arnaud, along with a team of veterinarians, also check the overall health of a captured animal before releasing it. They take blood and poop samples.

MOMS AND BABIES

Very little was known about mother and baby giant armadillos before Arnaud began tracking a female named Isabelle. After tagging her in 2013, Isabelle had a baby named Alex. "Through Alex, we learned a lot about giant armadillo **parental** care," says Arnaud. He soon found out that armadillos are great moms.

Arnaud checking camera footage for signs of giant armadillos

Before Alex, scientists thought that armadillo moms spent around six months with their babies. However, Isabelle lived with Alex for over a year. And she remained close to him up until his death in 2015 at age two. Arnaud believes a puma killed Alex. "I was truly very sad," he says. Female giant armadillos only have one baby every three years. So every baby is precious.

Alex was the first baby armadillo that scientists had ever studied.

LOOK CLOSER

Arnaud also tagged baby Alex with a transmitter. This enabled him to track the little armadillo's movement.

AT RISK

While pumas and jaguars pose a risk to giant armadillos, people pose an even greater risk. Cars often strike and kill giant armadillos when the creatures are trying to cross roads. And people hunt the animals for food and for their huge claws. Some local people in the Pantanal even kill the giants because they believe they're bad luck.

This armadillo was hit and killed by a car.

However, the biggest threat to giant armadillos is **habitat** loss. Farmers and ranchers are taking over the animal's homeland. As people move in, the shy animals are pushed out.

Arnaud also thinks that **ignorance** is a big problem. "The general public does not know this species exists . . . or believe it is already **extinct**," says Arnaud. "How can you care about a species you don't even know you have?" That's why Arnaud is devoted to educating people about these amazing armadillos.

LOOK CLOSER

Scientists think that the giant armadillo population has fallen by at least 30 percent in the last 25 years.

Arnaud at work

ARMADILLO CONSERVATION

In 2010, Arnaud started what is now called the Giant Armadillo **Conservation** Program (GACP) with a team of other scientists. The GACP research site is located at a ranch in the heart of the Pantanal. There, Arnaud and his team figure out how best to study the animals. They also monitor where the animals live, their health and behavior, and the threats they're facing. All of the information is stored in a database to help people better understand and, hopefully, save giant armadillos.

Giant Armadillo Conservation Program team members

Arnaud says one of the most important goals of the program is education. He and his team visit local schools to teach children about armadillos. The GACP team also talks with farmers and landowners about the big animals and why they deserve protection. "More and more people are now aware of the species' **existence** and the important role they play as ecosystem engineers," says Arnaud.

Arnaud teaching children about armadillos

LOOK CLOSER

The GACP team studies other kinds of armadillos, as well as tamanduas and other anteaters. The project has recently expanded to the Cerrado and Atlantic Forest biomes in Brazil to save wildlife there.

MUCH TO LEARN

It's not easy studying animals in the wild. Arnaud often works long hours alone in the hot sun surrounded by biting insects. But he truly loves giant armadillos and the other animals he studies. "It is these strong emotions and connection to our study animals that give us the drive and the strength to work so hard on behalf of their conservation," says Arnaud.

One of the giant armadillos Arnaud continues to follow is Isabelle. He was thrilled when he learned that she had a new baby. "Many questions still remain," says Arnaud. And the new baby will help the team learn even more about these amazing animals and how best to protect them. "The story of the giant armadillo is still being written, but we're discovering more day by day, week by week, and month by month," says a hopeful Arnaud.

LOOK CLOSER

Arnaud and his team are fighting to create protected areas of land for giant armadillos in the Cerrado in Brazil.

FAST FACTS

GIANT ARMADILLOS

Scientific Name
Priodontes maximus

Physical Description
Dark brown with a lighter band running along the animal's sides

Size
Up to 5 feet (1.5 m) long

Weight
Up to 110 pounds (50 kg)

Main Diet
Termites and ants

Habitat
South America

Life Span
Up to 20 years

DID YOU KNOW?

- A giant armadillo's teeth grow throughout its life.
- An adult giant armadillo sleeps about 18 hours a day.
- Before giving birth, a mother giant armadillo seals up her burrow to protect her newborn baby from predators. For the first 6 months of her baby's life, she closes the burrow each time she leaves it to feed.
- Between a giant armadillo's scutes are tiny, sand-colored hairs.

GLOSSARY

armored (AHR-murd) covered in a hard, protective layer or shell

biomes (BYE-ohmz) large, complex communities of living things in a region, such as a forest

burrow (BUR-oh) a hole or tunnel dug by an animal to live in

carapace (KAIR-uh-pase) a hard, shell-like covering

conservation (kahn-sur-VAY-shuhn) the protection of wildlife and natural resources

devour (di-VOUR) to eat something quickly

ecosystem (EE-koh-sis-tuhm) a community of animals and plants that depend on one another to live

engineers (en-juh-NEERZ) workers who design and construct buildings or structures

existence (ig-ZIS-tuhns) the state of being alive

extinct (ik-STINGKT) no longer found alive

field biologist (FEELD bye-AH-luh-jist) a scientist who studies living things outside in nature

glimpse (GLIMPS) a brief look at something

habitat (HAB-ih-tat) the natural home of an animal or plant

ignorance (IG-nur-uhns) lack of education or information

insectivores (in-SEK-tih-voorz) animals that feed mainly on insects

mammals (MAM-uhlz) warm-blooded animals that have hair or fur and nurse their babies

nocturnal (nahk-TUR-nuhl) active mainly at night

parental (puh-REN-tuhl) relating to a parent

predators (PRED-uh-turz) animals that hunt and kill other animals for food

scutes (SKOOTS) protective bony plates just under an animal's skin

species (SPEE-sheez) a group of similar animals that can reproduce

sprint (SPRINT) to run fast for a short distance

transmitter (trans-MIT-er) a device that sends out radio waves to help scientists track an animal

READ MORE

Jango-Cohen, Judith. *Armadillos*. Tarrytown, NY: Benchmark Books, 2004.

Llanas, Sheila Griffin. *Armadillos*. Minneapolis: ABDO, 2013.

Potts, Steve. *The Armadillo*. Mankato, MN: Caption Books, 1998.

LEARN MORE ONLINE

The Giant Armadillo Conservation Program
https://www.giantarmadillo.org

The Minnesota Zoo: Demystifying the Giant Armadillo
http://mnzoo.org/blog/demystifying-giant-armadillo

Mongabay: Armored Giant Turns Out to Be Vital Ecosystem Engineer
https://news.mongabay.com/2013/10/armored-giant-turns-out-to-be-vital-ecosystem-engineer

National Geographic—Diggers in the Dark: Discovering Giant Armadillos in Brazil's Pantanal
https://blog.nationalgeographic.org/2014/05/07/diggers-in-the-dark-discovering-giant-armadillos-in-brazils-pantanal

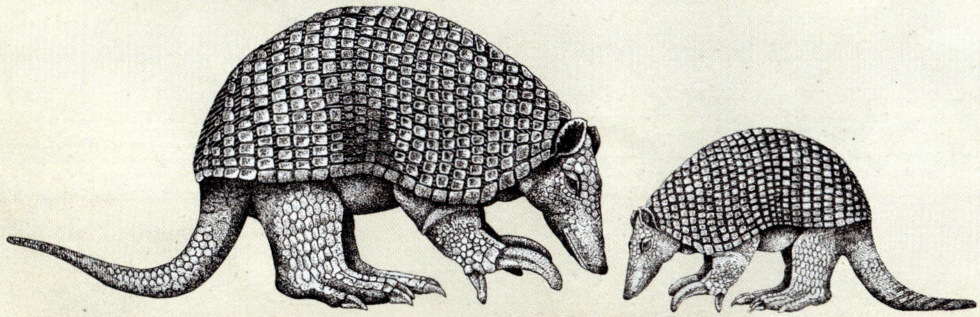

INDEX

ABOUT THE AUTHOR

Joyce Markovics has written more than 150 books for young readers. She's wild about rare and unusual animals and is passionate about preservation. Joyce lives in an old house along the Hudson River in Ossining, New York. She would like to thank Arnaud Desbiez for his contribution to this book and for being a champion of conservation.